Awesome, Disgusting Science

GROSS SCIENCE OF FUNGUS AND MOLD

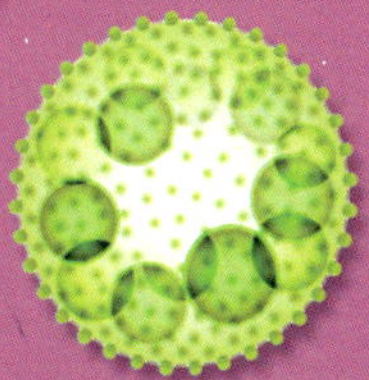

Stephanie Bearce

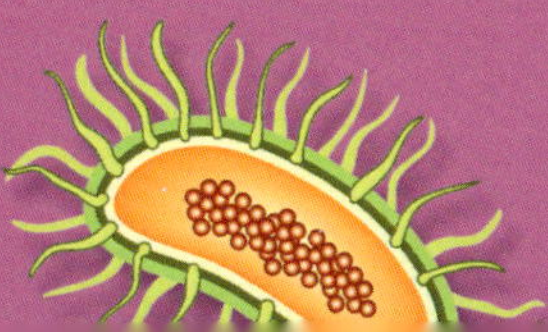

BLACK RABBIT BOOKS

Hi Jinx is published by Black Rabbit Books
P.O. Box 227, Mankato, Minnesota, 56002.
www.blackrabbitbooks.com
Copyright © 2026 Black Rabbit Books

Alissa Thielges, editor; Jason Knudson, designer and photo researcher

All rights reserved. No part of this book may be reproduced in any form without written permission from the publisher.

Library of Congress Cataloging-in-Publication Data
Names: Bearce, Stephanie author
Title: Gross science of fungus and mold / by Stephanie Bearce.
Description: Mankato, MN : Black Rabbit Books, [2026] | Series: Awesome, disgusting science | Includes bibliographical references and index. | Audience: Ages 8-12 | Audience: Grades 4-6
Identifiers: LCCN 2025017507 (print) | LCCN 2025017508 (ebook) | ISBN 9781645824930 library binding | ISBN 9781645824992 paperback | ISBN 9781645825050 ebook
Subjects: LCSH: Fungi—Juvenile literature | Molds (Fungi)—Juvenile literature
Classification: LCC QK603.5 .B43 2026 (print) | LCC QK603.5 (ebook) | DDC 579.5—dc23/eng/20250804
LC record available at https://lccn.loc.gov/2025017507
LC ebook record available at https://lccn.loc.gov/2025017508

Printed in the United States of America.

Image Credits

Adobe Stock/AndyPhoton, 17, Artinun, 16, barber, 19, PixelNexusArt, 16, Rodica, 16; Freepik/AI Image Creator, cover, 1, brgfx, cover, 1, 18, freepik, cover, 1, macrovector, cover, 1, 3, 8, 9, 18, 21, officialtrtasfiq, 11, 23, pikisuperstar, 9, smizurova, 4, 5, 12, 13; Shutterstock/Anna Hoychuk, 12–13, Astrid Gast, 12, Bertello Fabiola, 15, Chitaika, 8, Cholpan, 19, Elena Istomina, 7, Ezume Images, 4, Fahroni, 5, Igor Vitkovskiy, 16–17, Inamiqu, 7, Khazanova, 13, Komsan Loonprom, 8, Kreminska, 12, Memo Angeles, 20, Nataliia Dorosh, 6, 7, NIkita Oskolkov, 15, Paulo Cesar Ayres, 11, Petar B photography, 10, PRILL, 2, 3, 6, 7, Rani Restu Irianti, 14, Saskia B, 19, shurkin_son, 4, 11, Viktor Sergeevich, 15, Yuliia Gornostaieva, 12.

Every effort has been made to contact copyright holders for material reproduced in this book. Any omissions will be rectified in subsequent printings if notice is given to the publisher.

CONTENTS

CHAPTER 1
Putting the Fun in Fungi...5

CHAPTER 2
The Experiments......6

CHAPTER 3
Get in on the Hi Jinx...20

Other Resources..........22

Chapter 1
PUTTING THE FUN IN FUNGI

Stinky sneakers? Take a sniff. P-U! That awful smell might not be your socks. It could be mold! Mold is a kind of fungus. It loves dark, damp places.

Both fungi and mold can be deadly. But they can also be helpful. Read on to learn all about the weird science of these **organisms**.

Penicillin is made from mold. It helps treat bad bacteria.

Chapter 2
THE EXPERIMENTS

Drywall Doom

Black mold loves wet drywall. What happens when people breathe it in? Scientists found out. The results? Gross and scary! The **spores** hurt the lungs. Breathing was hard. Some lungs even bled!

To fight back, scientists tested a special shot. It was made of **antibodies**. It helped heal the lungs. Doctors hope to treat people with it.

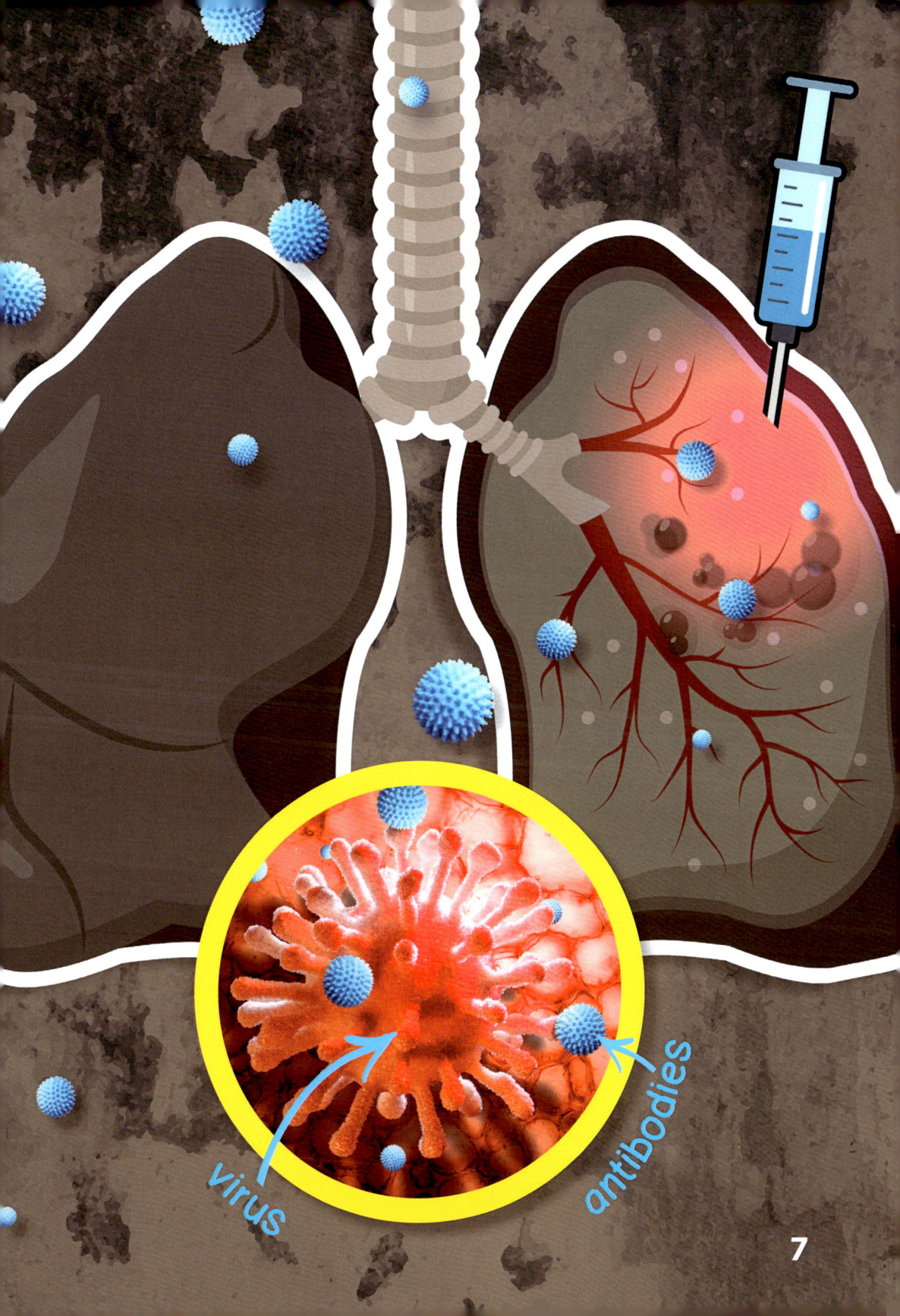

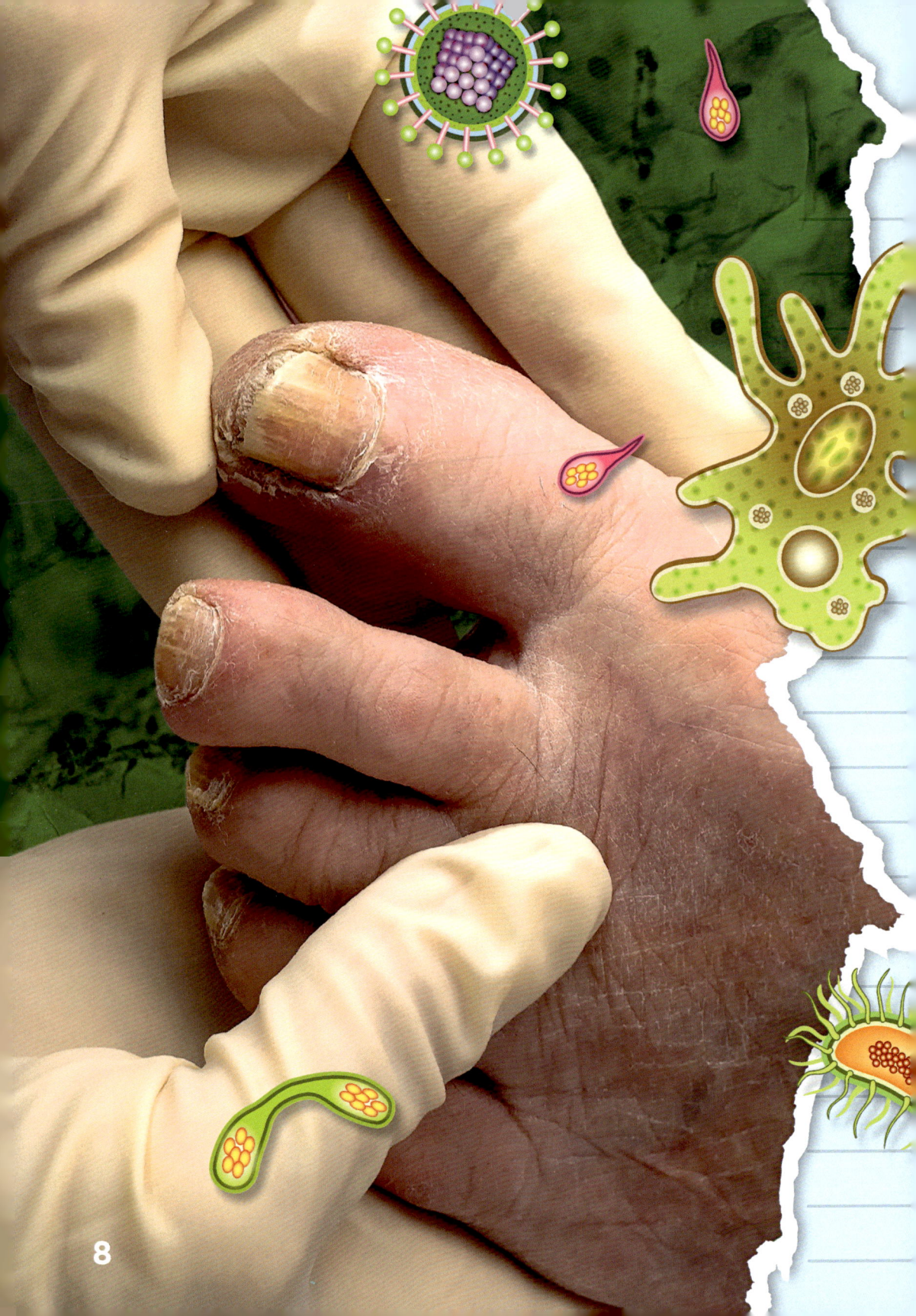

Fungus Among Us

Strength is usually awesome. But not when it comes to fungal **infections**! They laugh at any drug thrown their way. The illness spreads over hospitals. It grows in blood, organs, and even skin. It's like a total body takeover!

Scientists are on the case. They are cooking up new super-drugs. They hope the drugs will put these terrible infections in their place.

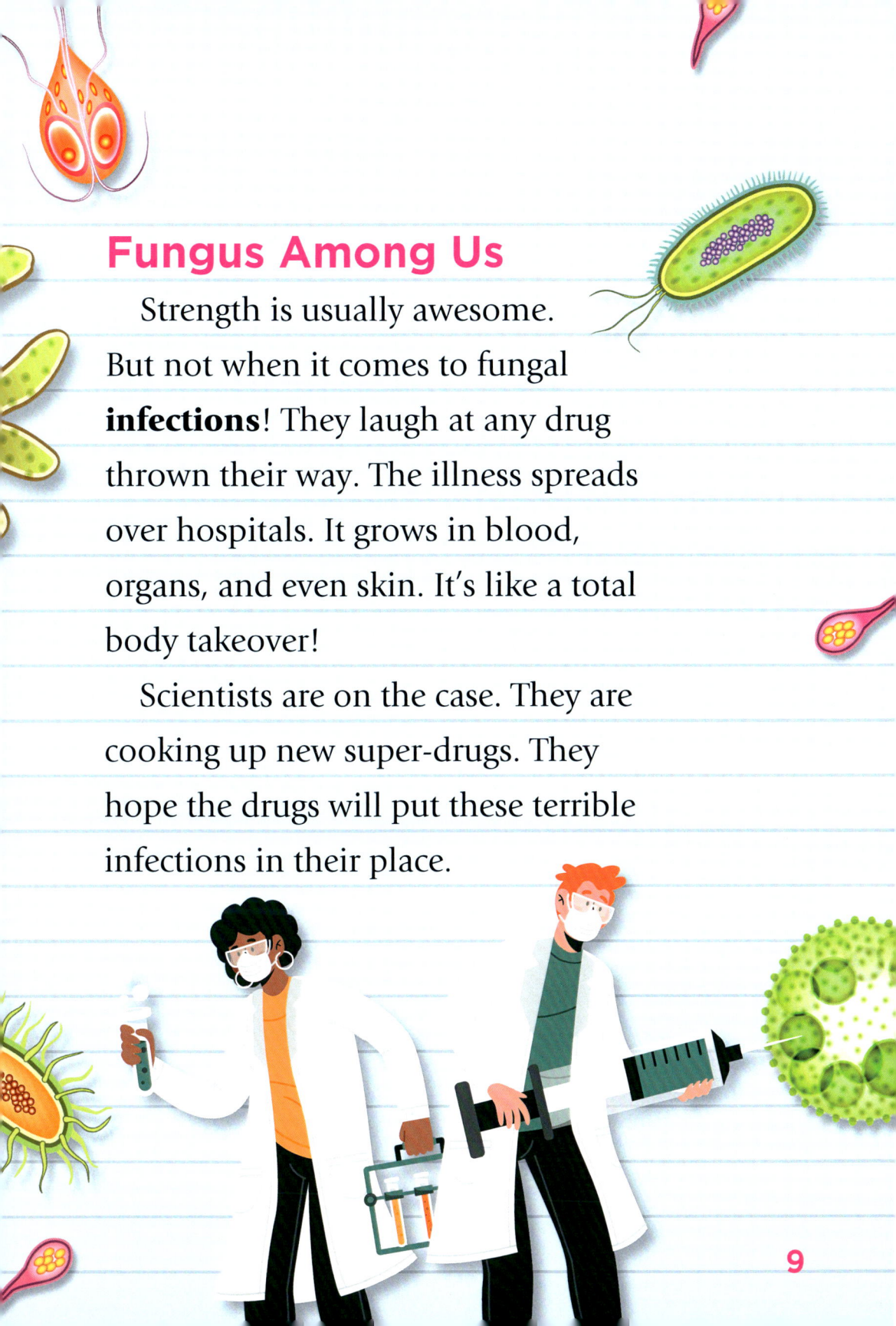

Brain Control

Imagine a fungus that creates zombies. It's real! The fungus takes over the brains of ants and wasps. The insects kill themselves. This helps the fungus spread to others.

Scientists study this fungus. They learned it may affect butterflies and roaches. But don't worry. This fungus can't spread to humans. Phew, that's a relief!

Some fungi glow in the dark. This is called "bioluminescence." Spooky cool!

moldy bread

moldy strawberries

moldy cheese

12

Munchy Mold

Fungi aren't all bad. Some help **ferment** tasty foods. Bread and soft cheeses? Yep, thank fungi for those. Soy sauce? You bet!

Scientists often poke around at moldy food. They even test how fast it grows. Try that excuse next time. "No mom. I didn't forget to clean my lunchbox. That mold is for science!"

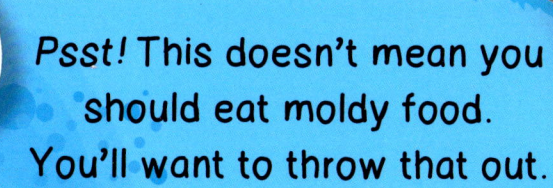

Psst! This doesn't mean you should eat moldy food. You'll want to throw that out.

Dumpster Delish

Garbage for dinner? It's possible with fungus. Scientists are using fungus to turn waste into food. The fungus breaks down **compost** and turns it into supper! Scientists say the food tastes like cheese bread. Fungus might help feed a hungry world. Would you give it a try?

This orange mold is called *Neurospora*. It is served in fancy restaurants.

Fungi Fashion

Think eating fungus is weird? Try wearing it! Mushrooms are a type of fungi. They have stringy fibers called mycelium. Scientists use this fiber to make clothes. They have made shirts, bags, and shoes. The fibers are a great swap for cloth and leather. Got a hole in your shirt? No problem. The fungus can regrow and repair it.

Mushroom Mansion

How about a house made of fungi? It's possible! Scientists are experimenting with fast-growing fungi as building materials. Imagine bricks made from mushrooms. You can grow walls in any shape you want. Got a hole in the wall? The fungi could seal it up. Don't worry—these fungi aren't dangerous. It's totally safe to live in a mushroom mansion.

Fungi in space? Scientists are exploring this. It could help future astronauts build homes on Mars.

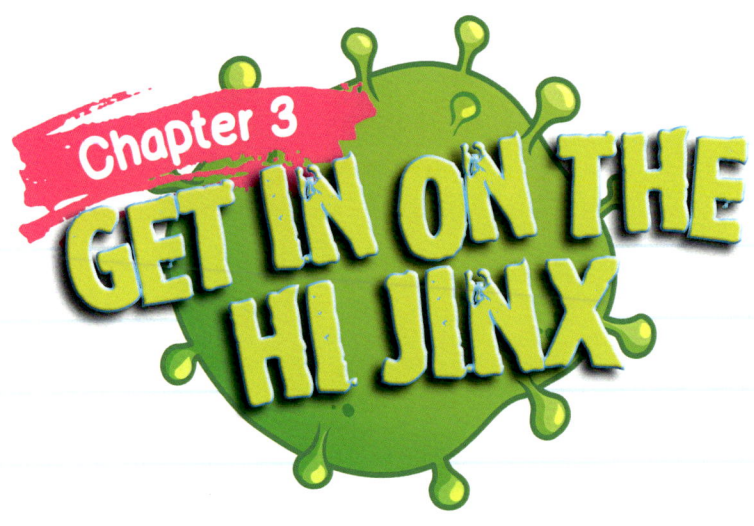

Chapter 3
GET IN ON THE HI JINX

Do you like poking around in the dirt? Then being a **mycologist** might be your dream job! They are like fungus detectives. They find new ways it can help humans. To get this job, study biology and ecology. Then learn in a lab or in the field. Collect samples and stay curious. Someday, you might find a new way to use fungi!

Take It One Step More

1. Imagine you are a scientist. What kind of fungus would you want to study?

2. Do you think fungi could survive on another planet?

3. If you could invent a new way to use fungi, what would it be?

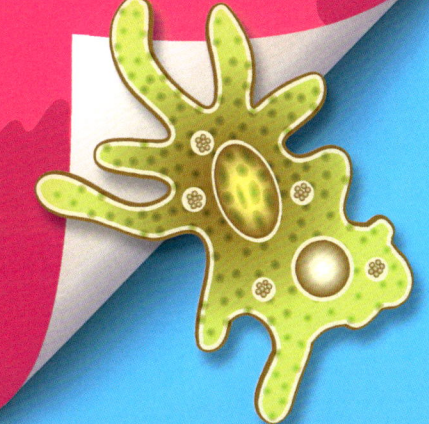

GLOSSARY

antibody (AN-ti-bod-ee)—a substance produced by the body to fight disease

compost (KOM-pohst)—decayed organic material, such as leaves and grass, used to improve soil

ferment (fer-MENT)—to go through a chemical change that results in the production of alcohol and gas

infection (in-FEK-shun)—a disease caused by germs that enter the body

mycologist (mahy-KOL-uh-jist)—a scientist who studies fungi

organism (AWR-guh-niz-uhm)—an individual living thing

spore (SPAWR)—a tiny particle fungi use to reproduce

LEARN MORE

BOOKS

Boddy, Lynne. *Humongous Fungus*. New York: DK Publishing, 2021.

Kroe, Kathryn. *What Are Fungi and Molds?* New York: Cavendish Square Publishing, 2023.

Loh-Hagan, Virginia. *Plants and Fungi*. Ann Arbor, MI: Cherry Lake Publishing, 2022.

WEBSITES

Fungi
www.ducksters.com/science/biology/fungi.php

The Fungi in Your Future: Mushroom Leather, Furniture, and More
thekidshouldseethis.com/post/the-fungi-in-your-future-mushroom-leather-furniture-and-more

INDEX

B

black mold, 6

building, 18

C

clothes, 17

F

food, 13, 14

H

hospitals, 9

I

infections, 9

insects, 10

J

jobs, 20

M

medicines, 5, 6, 9

mushrooms, 17, 18

S

smells, 5